Living with Grace has so many layers within it that I don't know why you wouldn't want to read it!

As an animal communicator I receive calls to help humans and animals communicate with one another in difficult life situations, end of life being one of them. It's very difficult to give the human information I know they don't want to hear — but they need to hear it. What I share comes directly from their animal, in order to bridge the gap of knowing and trusting what they need to do with the animal for them to let go and do the next step — whether it's holding on when you want to let go or it's letting go when you want to hold on. The animal is guiding them. There is comfort in that. It is a process between all parties, instead of one sided.

Sometimes the animal needs us to hang on just a little bit longer, for them to do their part in their life. Grace was an example of this.

Dying is an active process; if the animal was ready to die, she/he would have already left their body. We are here to learn to give grace to the dying process an be present with each other, so the death is more complete for all involved.

If you're someone who wants to go deeper, this is the book for you. Grace shows that it is not about getting it right in this moment. She shows that giving yourself space and time to come to your own Knowing is what's important. As Grace said quite adamantly to me during a session, "Give your own self some grace in this life!" This is what *Living with Grace* does — it meets you at your level and goes as deep as you want to take yourself.

Grace says, "Why would you not want to buy this book?" I second that!

Kristen Scanlon
Animal Communicator

What better way to get advice about life, and its meaning, than from the wisdom of a cat? In her book, *Living with Grace*, Marita Rahlenbeck shares her experience with her cat Grace, extracting the eloquence of Grace's message for us all. It is a beautiful exploration, whether you're cat-centric or not.

Carole Hyder
President, Caring for Cats
Author: *Wind & Water: Your Personal Feng Shui Journey*

I witnessed the story between Grace and Marita unfold while I went through the same thing with my dog. I had no idea how to deal with the loss of a pet — or any family member for that matter — so having someone I could relate to was very helpful.

Now, having read *Living with Grace*, I have far more understanding of how my Bootsie and her sisters, Tilly and Dixie, called to me. I understand the energy connection I have with them.

The book would have been perfect during my time of loss. Anyone going through the loss of a pet or loved one will benefit from reading it. I can honestly recommend the book to anyone grieving a pet...or during the process before hand.

Jackie Holdridge

Marita Rahlenbeck has written an endearing and memorable book, *Living with Grace*. The book is co-written by Marita's beloved cat, Grace. Marita and Grace will touch your heart and your soul with their message, openness, and profound wisdom. I found myself unable to put the book down, and wanting to know more about Grace the cat and grace, the state of being.

The book introduced me to aromatherapy and animal communication. At the same time it reinforced my belief in all things good and the essential need for grace in our lives. *Living with Grace* is a book for all ages; those who want to open their lives to new perspectives, those who are looking for a quick and immersing read, and of course, animal lovers.

L Davis

Living with Grace is an amazing account of the journey of two souls in Love and Understanding, from a human and four-legged friend perspective. This allows the reader to have greater introspection of their relationship with their own four-legged friends, and to learn how to serve each other with love. Journey On!

All things with Love,
Mary Gillen

Living with Grace is a thoughtful and thought provoking book based on the love between a cat, Grace, and her owner, Marita. The book reminds us that despite past experiences, all living beings are worthy of unconditional love and are capable of loving in return. It gives hope by showing that we learn to love by receiving and accepting love. It invites the reader to examine their own relationships in that same context; how we give love, how we receive love and how we lovingly say goodbye. A heartwarming, graced-filled read.

Much love,
Sandy CJ

I am deeply touched by your beautiful book, *Living with Grace*. Bless you and Grace, for sharing your love with the world.

As you so well stated, this is a story of love, allowance, grace, commitment, and how to weave in beautiful colors to the tapestry of our life. This is an inspirational must read for all ages!

Isabelle Giroday
Award winning author

Living with *Grace*

*A Story of Love
and Healing,
Leaving Paw Prints
on the Heart*

MARITA RAHLENBECK

Dedication

This book is lovingly dedicated to my daughter, Elizabeth. Without your persistent desire to look at kitties long ago, I would never have opened my heart to loving a cat. And, because you will always hold my heart.

I also dedicate this book to both Grace and Aria for choosing me as their human to join them on their life journey. Life would be uneventful without you.

And lastly, to all the men and women who work in animal rescue and foster care, and to those who adopt these animals. The world needs more of you!

Cast of Characters

Angela - Grace's veterinarian
Kristen - Animal communicator extraordinaire
Marita - Author and human mama to Aria and Grace

Grace - The subject of this book, co-author, and mama to Aria

Aria - The kitten I found in a sea of kittens

Ginger - My first cat, a regal feline queen

Your Turn: Throughout the book I will provide opportunities for you to look within and reflect on the topic and questions I have presented. You will find these at the end of each chapter and they will always start with the words, "Your Turn."

Interspersed throughout the book you will find quotes by Grace and me. In order for you to identify the speaker, Grace's quotes will have a cat symbol and mine will have a butterfly symbol.

 Marita

 Grace

Grace Notes:

Nourishment For Your Soul

Delivered direct to your inbox

Let Grace Notes provide you with:

- Encouragement to inspire confident action with renewed belief
- Nourishment for your Soul with healing words
- Inspiration to empower you and give you strength

Sign up for your Grace Notes here:

www.GraceNotesWithGrace.com

Table of Contents

Prologue

I once had a cat named Grace. Grace was a lovely animal who unfortunately suffered many abuses which left her unable to trust and accept love. Despite her past she taught me about grace, its importance, and how to find grace as she lived into her name. Grace ultimately let go into death and now we spend time together in a different manner.

My book, *Living with Grace,* is a compilation of my writing, Grace's writing, the Journey to Grace, and transcripts from my meetings with Grace and our animal communicator. The purpose of the meetings and writings was to bring Grace's words forward. From that, I offer opportunities for you to consider where, or how, you can access and bring Grace into your life.

During Grace's illness, I came to rely on our professional animal communicator who served as the bridge for Grace and me. The communicator was Kristen Scanlon. We both depended heavily on Kristen during this time. You'll meet her later in our story.

While our story takes you on a roller coaster of a myriad of emotions, Grace and I would like for you to come away with a new perspective on life, death, animals, spiritual gifts, grace, and how you have the power to impact your own life, and that of your pet(s) in a more active and inclusive way.

A friend recently provided me with some deep and profound insight into Grace and her behavior. Grace allowed herself time and space. She quietly and patiently observed my interactions with her kitten, Aria, and I believe this was her way of assessing trustworthiness.

My friend said: "When Marita traveled I would check on Grace and her kitten, Aria. Aria would immediately engage in play. Grace, on the other hand, would be off to my other side, far enough away to observe. She gave herself space and time to watch her kitten play. I gently tried to engage Grace with a red shoelace, tossing it towards her and pulling it ever so slowly away from her. She watched the shoelace intently. I made several attempts to engage Grace and just when I thought perhaps another day she will play, Grace would reach toward the shoelace and play! We were making progress.

Whenever I visited I repeated the playtime process. Each time Grace would eventually play with the red shoelace. Grace's process was to observe, keep her distance, and with time, play. She needed to, 'get ready' to play."

Grace's Foreword

This is my story.

This is our story.

This is a story of how my human loved me despite my imperfections. She loved me through the veil. Her selflessness brought me to acceptance and wholeness. This human wants to make a difference and I want to share Divine wisdom with you; together we are bringing you *Grace*. We are bringing you grace.

I broke through all of the pain, distrust, and trauma with her help.

If I had not trusted, moved through my fear, and dealt with my own trauma, fear, and pain (trust me that was not easy),

I would not be writing this book.

I would not have a voice.

I would have stayed quiet.

I want to use my voice for many reasons:

- So you can learn from me and your life can be easier.

- So you can move through the fear and the letting go.

🐾 So you can allow people to help you in your struggles.

🐾 So you can move through, be grounded and rooted in grace, and become an authentic version of your highest best self.

You don't need to die to get there. Let my death be the sacrifice, so you can live with more grace, joy, and bright colored threads in your tapestry. Surround yourself with people who bring you bright colored threads. Let others support you. Let them serve you. And then, you support others who need *you*. Bring your bright colored threads to their tapestry. We are all in the business of creating a tapestry with one another. We all wish for bright colors. Trust in yourself more.

Thank you for wishing to read the words my human and I have put together for you.

Breathe.

For in that breath there is grace.

In that breath there is space.

And in that space there is the Divine.

...In that space there is the Divine...

Grace 🐾

CHAPTER I

Time and Tears

They say time heals all wounds. This is true.

Guilt was big, overwhelming, and heavy. All the self-doubt would not stop rearing its loud, big, screaming head.

First the guilt was around, "Why didn't I see it?!" then to, "What should I have done that I didn't do?"

There was guilt around very specific happenings — in the moment I found them curious and indeed delightful — but maybe it was a sign? Maybe it had been a sign? Maybe it was an attempt to say, "Hey, stop. See me, hear me Mama. I'm hurting deep inside my body and want you to know; I want you to help me!"

What if?

What if?!!!

Even now, remembering, the emotion wells up and catches in my throat.

This is why I believe so strongly in animal communication. Kristen could have been Grace's voice had I circled her into our lives in a larger way earlier.

Here is a story to ease your curiosity and make this more concrete:

One morning, before I knew how sick Grace was, I opened the door and she ran up just four steps, stopped, turned, looked right at me and gave one distinct, "*Meow.*"

At the time, I was ever so delighted to hear her voice for she rarely used it. Now, I look back and wonder if it was her attempt to tell me she was hurting.

At the time of death, for pets and humans alike, there is a large cauldron of emotions amplified completely out of proportion and control. I began writing daily as a way to process Grace's death, my guilt, and the myriad emotions I felt.

As I look over my writing, page after page,
I see words such as:

Pain

Loss

Guilty

Sorrow

Regret

Overwhelming guilt

Frustration

Sadness

I wanted a do-over.

And then Grace steps in to show the other side — the side I struggled to get to:

Acceptance

Love

Forgiveness

Intention

Reconciliation

Reconciliation from within myself because Grace, of course, had already forgiven me; indeed from her perspective there was nothing to forgive.

They say time heals all wounds. This is true.

Time

Forgiveness

Tapping

Emoting

Tears

Essential oils

Silence

Focusing attention on Aria

Telling the story for myself and then discovering how moving and impactful it was to others. This was healing as well.

Grace from the Other Side

My relationship with Grace changed once she made her transition. She became my guide and my teacher. Her profound wisdom was impactful. I began meeting her with my journal each morning. I looked forward to getting out of bed to spend time with her each day.

I was astonished and delighted that I could hear Grace speaking to me immediately after her crossing. She told me what she wanted me to do, where I should go, and what she preferred. When I thought of a question, I began *seeing* the answer. I never spoke the question out loud or took time to think of an answer. The answer just appeared before me. This was cat magic working in my being-ness and life! I found it very exciting!

I'm always up for a challenge, especially one as unique as writing with a cat that is on the other side of the veil. I looked my grief in the eye and dared it to bring me down.

To be asked to write a book with a cat is not your everyday, average occurrence. Yet, after Grace died I found myself meeting her every morning for many months. It was a way for me to process her loss and connect, not only with her, but with the Divine Wisdom she now had access to and so willingly shared with me. I looked forward to getting up in the morning and sitting with my journal, special pens, and Grace. I was excited to see what would come through during our time together each day.

At some point the opportunity for publishing a book became more tangible. The idea went from writing a book "someday" while thinking to myself, "Grace! I don't know *how* to write a book" to, "I'm writing a book; *we* are writing a book!"

I knew if I hesitated and thought about it, rather than feeling my way through the process, I would choose the easy path. Of course the easy path meant not writing the book, which means you wouldn't be holding it in your hands right now. More importantly, you would not get to meet my sweet Grace and receive all the wisdom she wants to share with you.

Her journey is the choice of the hero. The hero who trusted her knowing, trusted the process, and took one paw step forward with blind faith. And repeated it; again and again.

Did I know how to write a book? The human in me did not, but my soul sure did. When I let go of the human me and allowed my soul to take the steering wheel, it became easy.

Your Turn

Where in life are you being asked to write with grace? Is there something you have wanted to say to someone but have been afraid to voice? Is it too painful to say aloud? Are you unsure or afraid of how it will be received? Maybe you've been thinking about someone you've lost touch with? Or someone who walked out of your life (or you theirs) and now it's up to one of you to take the first step in healing that relationship.

It's your turn to allow and trust. Write that person a letter or short note. Call upon grace for assurance. It can be as easy as taking a deep breath and beginning. Begin. Feel your true self and the message you want to convey. See their bright eyes and big smile. Feel their openness to you and your words.

Blessings await you on the other side of this action. Forgiveness is on the other side as well. You may even feel your shoulders loosen and notice your jaw is no longer tight.

Take the bold action. Write with grace.

Write With Grace

CHAPTER II

Starting with the End

The closer I came to death, the more present I felt.

Icalled to make the appointment first thing in the morning. The schedule was full, but when the staff heard it was about Grace they squeezed her in at 6:00 PM, the end of their long day. For me, that may as well have been six months out. The moments crawled by. It was a long, painful, emotional day.

Grace was lethargic. I laid her near lifeless body on a towel by the patio door. All day she did not move. She barely breathed. She went in and out of her body as its systems shut down. She was warm and then she was cold. Eventually I put a towel over her like a blanket. I surrounded her with crystals. Aria looked bewildered and cried. All my attention went to Grace and getting both of us through to our six o'clock appointment.

My Aria was neglected.

My poor Grace was readying herself to transition fully. As her body weakened, her Spirit grew strong. All I could see was her body becoming increasingly lifeless. This witness was the most difficult thing I have ever done. I gave my Grace "grace," and allowed her to have a say in this one last thing. This important one last thing.

Shortly after her death Grace said to me, "I came to learn love and it was hard. Hard because it took years to find and I had given up. People were mean and unkind to me. People don't think of cats as valuable, as sentient beings with heart, emotions and love. Cats are thought of as things and we are misunderstood.

My message is to teach you our value — our value is found when we are met by our human — an evolved human who understands we are the same, yet different. My human, she understood, yet didn't. My human loved me like no other human did in that lifetime. I held back because of pain and distrust, and this impacted her ability to fully serve me and understand me. It was a time of learning for both of us. She is still in her humanness, so she sometimes hurts. But I am free. I am free of the frail, sick body, and all is forgiven. She did as I asked. I am at Peace and ready to serve from the other side."

Grace spoke to me and said, "I felt broken. The closer I came to death the more perfect I felt. Yet, the more perfect I felt, the more I didn't want to leave you. The more I let you love me, the more whole I felt. The more whole I felt, the more I could love and trust you. The more I allowed myself to love and trust you, the more I wanted not only to love and trust you, but I wanted to STAY. I wanted to STAY Mama. I wanted to STAY!

But my body couldn't. So I did my soul work so I could let go in peace. In Peace.

Thank you Mama for loving me so fully. So very fully.

I felt the fear of letting go. I finally had a reason to stay and then it was too late. By then I had only one choice and that was to let go into death. To let go of my dying body so I could be free. Yet, I was torn by the realization that I really did love you and that I could not stay because my body could not sustain itself. I finally knew and understood your love for me. And I wanted to be with you. I wanted to live in the body you called Grace and to spend time with you.

But it was too late, Mama. It was too late. Ultimately you helped me let go in a most holy manner. You helped me let go Mama. Now I am free and I get to spend time with you this way."

Your Turn

Grace's most poignant words bring up a lot of emotions. Reflect on what they bring up for you.

What fears are you facing? For you, what requires the act of letting go? Fear and letting go are part of the life journey toward wholeness.

Make a note of what scares you in your life — at this very moment. Then, take a few moments to visualize two scenes. Center yourself with a deep breath and allow the process to show you:

1. Your life in one, two, or twelve months when you give in to fear and continue to hang on.
2. Your life in one, two or twelve months when you choose to let go.

Compare and contrast. Yes, you're probably making something up. But maybe not! Your higher self knows what's at stake. Let yourself feel what the best course of action is. It's the ego which causes us to hold onto fear.

Now take those two stories and think about your life. Look back over your life and notice the difference between the scenes. Notice the feelings in your body and your heart.

Write With Grace

CHAPTER III

The Regal Queen

What does one say when a precious companion asks permission to return? 🦋

The first cat I owned was a regal queen named Ginger. She found her way into our home due to the constant pleas of my daughter who wanted to look at the cats at a local Petsmart. "Please, please, please, Mom!" she begged.

Ginger spoiled me, in that she embodied the attributes of a regal queen. Consequently, I was unprepared for how life with another cat, a complicated non-queen feline, could be.

In order for you to understand my journey with a cat named Grace, I need to explain the back story. The regal queen, Ginger, had become very frail and it was difficult for me to witness her suffering. She was sixteen-years-old when I made the difficult decision to let her go in order to relieve her pain.

Just as I finished bawling about having to put her down and declaring I would never have another cat — this was just too hard — I received a call from someone I did not know. A mutual friend of ours had asked this person to call me. She was a psychic. The majority of our conversation is a blur due to my heightened emotional state and because of a traumatic brain injury I suffered the previous year, but what I do remember her saying is, "Ginger would like to come back. With your permission, she'd like to come back to you." I'm sure my eyes opened wide. After all, I had just declared, "No more cats," which really meant no more heartbreak.

What does one say when a precious companion asks *permission* (how queen-like is that) to return? Of course I said yes, but with two conditions: I needed to be emotionally ready and it needed to be after an upcoming trip to China.

A few months later, my China trip behind me, I began opening up to the possibility of owning a cat once again.

Knowing that Ginger wanted to return to me, my biggest question was how would I find the "right" kitten in a sea of kittens? I asked myself this question repeatedly.

One afternoon, while stopped at a red light half a mile from home, I said out loud, "How am I supposed to find *you* among all the kittens? And then with my second sight, I saw a black cat as the light turned green. That's it. A black cat. Really? There are so many black cats! It was not much, but it was a start. Now what?

Allowing and Trusting

Grace's story will be revealed in these pages, but for now, know that by finding Aria, a kitten who embodies Ginger's soul, I found Grace.

From my perspective, Grace lived a short life in terms of how long a cat "should" live. But for her, it was the longest time she ever spent in a body. She's gone from our physical world, but lives more powerfully and boldly now than she ever has. I have partnered with Grace in order to give a voice to her wisdom through this book. This book is a synthesis of our writings together.

As I shared on social media during her illness and transition, many people reached out stating how much our story had impacted them. Grace's journey is one of healing. Indeed, she herself said, "My story is their healing."

Grace and I wish to bring meaning to her suffering by sharing our journey with you. She suffered from an unexplained and ultimately undiagnosed illness. It was during this time I started writing about her journey. Ten days later my Grace was gone.

It is our sincere and deepest desire that this book blesses you, shifts you, inspires you, and gives you a new love and appreciation for cats — indeed all animals.

Your Turn

Think of a time when you were asked to trust something far outside the box. What did you do? What was it like to trust and to be blindly guided? Did you ask for support or do it all alone? Did you share your journey along the way?

Write With Grace

CHAPTER IV

Traumatic Beginning

Sometimes we are so wrapped up in ourselves, we forget the other person needs a kind word, a moment of attention, or to be included.

Ruff Start Rescue had what I call a rigorous application process which included an in-home interview. The volunteer wanted to see my house and meet me in my environment.

One of the questions she asked was, "Where will the cat sleep?" I thought to myself what an odd and curious question. I remember pausing for a long while wondering if she was serious. When it appeared she was, I said, "Wherever she wants." That answer seemed to satisfy the volunteer so we moved on.

Keep in mind I really only had one reference point with cats and that was Ginger, my very regal queen-like and extremely well behaved cat.

It was winter and Grace had been hit by a car and left for dead at the side of the road. A woman drove by and felt compassion. As she looked in the rear view mirror, Grace raised her head. The woman went back, picked her up, took her to a vet, and Grace eventually found herself in a foster home with Ruff Start Rescue.

As a result of the collision, Grace had many injuries. Ultimately she lost an eye; she had a quirky tilt to her head and her gait was off a bit. She suffered a brain injury, yet did not consider that a big part of her story.

There was trauma which expressed itself as distrust of humans and terror of loud noises, especially diesel trucks. (It wasn't a car that hit her, it was a truck. This was deduced from observation and verified by her behavior.)

After the hospital and vets did what they could for her, she was in foster care to await the birth of her kittens. The name she was given by the rescue was Mama Kitty. Why? She had four kittens and was a surrogate mother to three other kittens who lost their mama.

My Grace was not, "Mama Kitty!!" When I heard her story, I heard "Grace." She felt like a Grace. She felt like *grace*.

After all, it sure felt as though she was saved by grace. Especially in her last weeks as she fully embodied the state of being of grace.

When she was told that her name was Grace, her response was, "That suits me." I don't think we knew the profoundness of my naming her Grace and her recognizing that the name suited her.

The first night Grace spent in my home she had full reign of the house, which of course meant the bedroom. As I was sleeping that first night, I was awakened when Grace jumped up on the bed. Suddenly, I was alarmed! She was using the bed as a litter box. As you can imagine, this did not go over well.

In the middle of the night I not only had extreme emotions, but I also had to strip the bed and do laundry.

This was our tumultuous beginning. Needless to say, our relationship had a very rocky start.

Over the course of the next few days it became clear that Grace was either acting out or she had never been trained to use a litter box?

Here was this beautiful traumatized cat that I knew in my core I was to share my time with, yet my pre-learning with cats did not entail anything to prepare me for this type of behavior. I had been extremely spoiled by Ginger.

The inappropriate peeing repeatedly landed Grace in my laundry room. She did not like the small room, but it was a room that was easily cleaned. It was getting so bad and I was so distraught, I was texting her former foster mama telling her I was considering returning Grace even though a part of me was screaming, "How do you return a living, breathing being knowing the chances of re-homing a traumatized, one eyed cat who doesn't use a litter box are extremely low?" That did not sit well with me, thus I found myself in a quandary.

Upon hearing my story, a friend of mine came over. She had such compassion for Grace locked up all by herself that she went to the laundry room, sat herself on the floor, and just talked and waited for Grace to engage.

Later, she timidly asked me if she could bring Grace upstairs. My answer was, "Yes, as long as we are on the deck."

There we found ourselves — two women, Grace, and her kitten Aria — outside on the deck. I was confused, and I'm sure Grace was too.

This is when I called Kristen asking for her help. Kristen's amazing, holistic gift and perspective truly helped me look at the situation with compassion. She gave me a bird's eye view and I could see more than just her undesirable behaviors, but the trauma that caused them. That afternoon began a new journey for my family, which included Kristen when I needed an interpreter.

All of this leads to the conversation about how trauma — emotional as well as physical — impacts behavior and how we may act out, without even realizing we are acting out. If we do not realize we are acting out, surely we cannot expect those around us to realize it either! When we understand the back story of someone else's past, their injury, or their story that then adversely triggers something within us or them, it helps us have more compassion. For example, if I am acting out and I realize it is because of something that's happened in the past, then I can have more compassion for myself and step away from personal judgment.

The same holds true when interacting with other individuals or animals. There is always a back-story. If we are allowed to peek inside that story, something unleashes in the heart. That something is compassion, grace, understanding, and acceptance. That is our goal — to accept, to understand, and to have grace and compassion for ourselves, others, and animals with whom we are in a relationship.

Toward the end of Grace's life when she was so sick, I could not fathom the idea of not having her by my side. I was brave and brought her into the bedroom. One morning toward the end of her short life, as she jumped off the bed she lost control of her bladder. This time though, I did have compassion and understanding.

She said, via Kristen of course, that she didn't understand what had happened. Her body was breaking down around her and she didn't understand. I went about the task of stripping the bed, all the while keeping an eye on her, loving her, and giving her grace.

We are called to give each other grace. Sometimes we are so wrapped up in ourselves we forget the other person needs a kind word, a moment of attention, or to be included. We forget they just need *grace*. Offer grace to others whenever possible.

Your Turn

——————◁◆▷——————

Have you recently felt as though no one hears you? Felt slighted or excluded? It may not be easy, but I encourage you to voice it to those that matter and those who will listen to you.

Have you ignored someone that needs desperately to be heard, seen, or included? Offer grace. Be inclusive and offer compassion.

We all need each other. It begins with you.

Write With Grace

CHAPTER V

Hattie the Black Cat

Miracles happen.

Miracles happen in the everydayness of our lives. The miracles happen when we are aware of signs, prompts, and premonitions. As we learn to become aware of the voice, the feeling, and the vision (indeed some people can see) we can practice the Art of Listening, the Art of Allowing, and learning how to act upon what we have received.

Previously I mentioned that I had asked Ginger how I was to find *the* kitten and she had shown me a black cat. Here is the story of how I found *that* black cat.

Across the street from my salon is a Petsmart that has animals available for adoption. On a December morning, being ahead of schedule for a haircut, I stopped in to satisfy my curiosity about what animals were available for adoption that day. There were only two cats that day. Both of them were black! One was extremely social, gregarious, and very delightful. The other was distant and stayed in the back of the kennel, yet she had an intriguing energy. I found her compelling. I went back and forth and gave them equal attention. The quiet, distant female eventually came closer. Later

I could touch her nose and paw. She impressed herself upon my heart. Her name was Hattie.

Aria had shown me a black cat and I followed my inner promptings. This was my introduction to the black cat.

I saw this experience as a puzzle piece and yet the puzzle, or as Grace will teach you, the tapestry, was far from complete. I didn't know how this thread within the tapestry would ultimately come to rest or what picture it would create.

About ten days later I found myself in Petsmart again. I was curious if Hattie was still there. As I walked toward the adoption area I noticed the door was propped open and there was lots of excitement and activity. I learned that someone was picking up their cat that day! A representative from Ruff Start Rescue was there as well. This is the piece that the black cat, Hattie, led me to. (She was still there by the way)

This representative told me about their rescue organization and their process of adoption. She assured me it was easy and that during the course of our conversation I had already completed the interview process.

This was all part of what I would later learn was the orchestration of Aria calling in her physical mama who would birth her, and me, her human mama.

Your Turn

Recall a time when you followed nudges and your own inner voice and found yourself in a most unexpected and delightful place.

What does it *feel* like just remembering? Where in your body do you feel it?

Take it a step further and write about it. I find when I write about experiences, I remember more details.

If your memory involves someone else, pick up the phone and give them a call. Or write them a note. Share the joy!

Write With Grace

CHAPTER VI

Animals Choosing their Humans

"That's the family I want. We need each other." ~ Aria

You may recall we found Ginger because of my daughter's incessant pleas to go to Petsmart. There were only two cats at the store, a male and a female; the female, who was about two-years-old, ended up coming home with us. We changed her name from Patches to Ginger. The beginning of my love affair with felines had begun.

Over time I realized Ginger had chosen *me*. I thought I'd bought the kitty for my daughter, but Ginger became my cat. She was a very regal queen who showed me the regal side of myself. She was a powerful healer kitty who helped in my healing energy sessions with individuals.

During my patient waiting to find a kitten with a rescue organization, I learned that the people who volunteer there, and do all their good and noble work, look at cats as a commodity. Over the months I'd been passed from one person to the next who understood their job was finding me a kitten. What was I looking for? A kitten. What color? I have no preference. What breed? I don't know; I will know her when I see her (feel her).

Looking at the tapestry of how all this weaves into grace and the wisdom of the animal kingdom, let's listen in on the conversation with Aria and how and why she chose me:

Grace said to me, "That was Aria. Aria went into your head. I didn't do that. Aria is a very powerful kitten. Aria's wisdom goes deep. Her magic goes even deeper. She is experiencing this life in a way that she's not experienced before. That cat can manifest like you'll never believe.

Aria pulled you into the relationship. She brought you to us. Aria did the whole thing before she was even on this planet. She picked you!"

When I asked why, the response was, "Why wouldn't she?!" It was a very humbling moment.

Just as we humans choose our parents and our life lessons, so do our animals. Why should their soul journey be different than ours? It's not. Aria wanted a direct path into my life. She didn't want to bounce around with uncertainty and trauma like Grace. She wanted stability. She chose me as her human mama and Grace as her mama. She saw and knew that Grace needed my grace, love, and compassion and that I needed hers.

Aria said, "That's the family I want. We need each other."

Aria choosing her own path was her grace for her mama, me, and herself. Aria chose Grace even before she knew her mama's name. She chose grace. She knew what her mama needed and set out to find me.

Meet Ernie

I am confident that animals choosing their humans is quite commonplace. While we would like to think we are the ones in control of this, we are not. We, and they, are looking for the right "soul connection."

I recently met a man on a walk who was patiently waiting for his very curious two-year-old puppy to be done being curious. The dog was a gorgeous long haired German Shepherd. It had the sweetest eyes and demeanor and was gracious to boot.

What I've learned about people is that they love being heard, but even more they love telling stories! The shepherd's owner opened up and told me how he came to have a dog. He had recently retired and his family suggested he get a companion.

He said to hold a puppy is the "kiss of death." I knew he didn't mean it the way it sounded. He meant it as his heart would cave and he would not be able to resist. While he did "look" for a puppy on two previous occasions, it was not until he held Ernie that his heart melted.

Now he takes Ernie for a walk three times a day.

Your Turn

Recall a time in your life when you recognized prompts or guidance like this. Did you act on them? If so, remember the experience. Recall the details. How did you feel? What was the outcome?

If you didn't and only recognized that you were indeed given prompts and guidance, how did you feel once you realized you'd made the "wrong" choice? Relive the experience and rewrite it.

If you are a pet lover, have you had the experience of being chosen by an animal? If so, sit and recall that experience. Learning from the past can give us a foundation to make better, wiser choices in the future.

Write With Grace

CHAPTER VII

Animal Communication

The beauty of Grace's life was the gift of presence.

Animals are our teachers, our companions, and our family. I have learned a lot about honoring animals as soul and allowing their voice to be heard. To do so is hugely important. I hope one day that having an animal communicator is as customary as having a dentist and favorite stylist. If you have pets, you should have an animal communicator to take the guesswork out of being their human.

I came to rely on a professional animal communicator to be the bridge for Grace and me. Over the course of Grace's illness and eventual death, both Grace and I relied heavily on Kristen Scanlon. She is my animal communicator of choice.

I don't pretend to know how the gift of such a talent is bestowed upon someone, just as I don't pretend to know how any of us come to realize, accept, and embrace our inherent spiritual gifts. What I appreciate most about Kristen is that she is not only the voice for the human to the animal or the animal to the human, but she brings in the Spiritual aspect of who we are as Beings. Because of her ability to bring in the higher vantage point, I have experienced the tremendous value this brings to a relationship.

The beauty of Grace's life, in my opinion and from my perspective, is that she was given the gift of "presence." I understood the importance of the Soul's Journey, so I made decisions from that vantage point around her life and death. With Kristen's help, Grace could give me comfort and guidance while helping me understand what I was witnessing.

Here's an example: when I witnessed Grace's body nearly lifeless it was extremely alarming. Kristen could tell me (could she see it? sense it? was Grace telling her this?) that Grace was going in and out of her body due to the pain. That was her conditioned response during unpleasant encounters and trauma. (It is also common for humans to do this as well.) Once I knew what I was seeing, I began to recognize this and it was no longer alarming. Knowing this I could better support her.

I believe that animal communication is truly an important part of a family dynamic, especially when there are behavioral issues that we cannot understand.

Communication with Grace

You may be wondering how Grace and I communicated with one another? How was I able to hear her and understand her?

That is something I cannot answer because I don't know. Once she transitioned it just happened. I did not need to try or force anything. It just happened.

Here is an example: In my neighborhood there is a small strip mall. On one end there is a resale shop with high end items and on the other end is a grocery store. A few days after her death, Grace was extremely insistent that I go to the resale shop. I acquiesced, fully expecting her to show me something I was to buy, yet I found nothing compelling. I did however notice some lovely note cards. As I was paying for the cards a song began to play. It was a song from my youth. When I heard the words I was surprised I didn't break down and cry a puddle. The song was Jim Croce's, *I'll Have to Say I Love You in a Song*. Wow! The swirl of emotions was huge. Now Grace could communicate with me in a way I didn't have to second guess myself.

As I walked to the grocery store, I felt my emotions bounce from sheer joy and love to grief and sadness. Then, moments later as I walked through the produce section the same song played *again*! The original version from 1973! I had not heard the song in a very long while and now in a span of fifteen minutes, I'd heard it twice. This is what animal communication can look like.

In the beginning when I would sit with Grace to write, I used an oracle deck as a prompt. It was effective. Once the flow and connection began, writing was effortless.

Many of Grace's "quotes" are from those writing prompts.

Your Turn

Have you ever used, or considered using, an animal communicator to help you understand a pet's behavior? If so, how has it helped you in your relationship with your animal? If not, why not?

I remember when it was first suggested to me years ago. I had no idea there was such a thing. Be open to trying new things, you never know where it might lead!

Write With Grace

CHAPTER VIII

Finding and Naming Grace

I was following energy and didn't know why.

I fell in love with Grace having never set eyes on her.

I found Grace less than two years before her last breath via a phone call about a pregnant cat who had been hit and left for dead on the side of the road. The resulting trauma and injury didn't enter my mind; there was a knowing deep inside me that knew this cat was my cat — as was one of her kittens.

Was I interested? I *felt* the answer. Yes, I was interested.

I waited. I was not in a hurry. I knew the right kitten would find its way to me. And I knew this hurt, pregnant cat was how I would find that kitten.

I did not know how much she would teach me or how wounded she was and how tumultuous that would initially make our relationship.

At some point Grace and I had a conversation about how she and Aria found me. I asked Grace, "Were you on a conscious intention of finding me or were you on a conscious intention of finding your way out of whatever your past had been?"

Grace answered, "When I am in this form, it is a path of finding you. But in my own body when I look at it in that moment, I was following energy and I didn't know why. I was being pushed somewhere. I was pushed to go left instead of right and it landed me in a place for another experience to happen. That's really what it was."

"Were you looking for an energetic signature or soul?" I asked.

"I was looking for soul family. I was looking for a soul connection. When you recognize someone you are meant to be friends with or partner with, or possibly you're meant to be in that person's life, you know it when you meet that person, that being, that place. You walk into a space and you know that is where you need to be in that moment. That's how it's recognized. Animals are so much better at doing that. We are usually connected in."

"You knew. You knew when you met me," Grace said.

This was the first photo I received of Grace. She still had both eyes. She looks as though she feels so sad in this photo.

"I knew before I met you," I answered.

And so I fell in love with Grace, without having ever set eyes on her. Indeed, it was a long time before I was sent a photo. She always hated having her photo taken.

What's In a Name?

Grace is a beautiful name isn't it? A beautiful word. I find it a difficult word to describe. For me, it's more of a feeling. I consulted the dictionary to help me with the word grace. Per www.dictionary.com this is what I found:

Elegance

Beauty

Favor

Goodwill

Pleasing

Mercy

Grace period

Unmerited favor

Excellence of divine origin

Moral strength

A blessing

Goddess of beauty

Grace note

I love the play on words with "grace." I can say I'm writing with Grace and yes, I can mean I *am* writing with Grace the cat and yes I am also writing with grace, the state of being. Throughout the book, when it is not clear, you may find yourself wondering which I am referring to. I have done this intentionally, as you are then allowed to interpret it in the moment.

Your turn

Is there an area in your life where you need to give yourself more grace? Or have you been looking for something and now realize you have been seeking grace? Has grace been hiding in plain view and suddenly you see it for what it is?

Take a few moments to reflect on these questions. Take any appropriate action.

Write With Grace

CHAPTER IX

The Tapestry

*Every paw step forward is part of that journey of living
with Grace.*

When Grace first explained the tapestry to me, this is what she said, "My story is a story of love. It's a path of love. Every step we take, every paw step forward, every choice, decision, thought, everything is part of that journey of living with grace. It doesn't stop when you die. It never stops. Ever. It's a continuum. Whether it's a paw step forward (paw steps are smaller than human steps), whether it's a little step or a big step; it doesn't matter. It's part of the journey. It's part of the tapestry."

Grace talked a lot about the tapestry of life. Tapestry is the whole picture; the art, the beauty of life, and what it is. It's "the whole." While we are *in* life, making each choice and going down the path that leads from one choice-point to another, those are known as the ***weft***, they are *us* making choices — good ones and poor ones, it doesn't really matter — they are the choices which take us on our life journey.

The ***warp,*** or stationary yarn, is soul and is always there. It is a rock of confident knowing. It is confident grace. It never doubts why it is here or what it came to learn. Our soul allows us our humanness and our choices. When we allow and ask, the soul answers. The warp, our divinity, is all we know before we incarnate into physical form. The warp is Divine perfection.

Once we come into the physical, the weft begins to create our picture. This is when our tapestry begins. We are now independent beings capable of making choices and going along the path our choices lead to. When we make a less than perfect choice, typically we were not listening to our higher knowing. If we are not listening and continue to make the same choice from habit rather than truth, life happens. Sometimes there's a catastrophe, injury, or accident. It is in those moments we are forced to go still, because at that point there is no choice.

When there has been injury, there is a corresponding recovery period. If it's a sprained wrist, broken toe, head injury, or something even more severe, there is always a corresponding healing that is necessary. Everything else, everything you thought was important before this incident, now becomes faded and its importance diminished. Those are the wefts of our tapestry. These wefts create the picture of our life.

Many times our wisdom and our deepest learning comes from these episodes. Where we go during these times is the dark night of the soul.

The dark night of the soul is a place of deep spiritual learning. It is a deep learning where everything else falls aside and the only thing that matters is healing, taking care of our body, and returning to a thriving, whole state.

When Grace lay on the side of the road, she didn't say, "Let me die." Instead she said, "Get me to the next place so I can get out of this, get me to the next place so I can get out of this, and to the next, and so on."

She knew at that juncture, that these choices, these decisions were being made on her behalf, but she had to trust because she had no strength. She didn't really have a say, unless she gave in to death. And yet, the fascinating thing is she didn't give in to death! How many times in her short life could she have relinquished and let go and gone to the other side? How many times could she have made that choice and taken that weft in the road that would have taken her home to wholeness. She feared death because she had not yet made peace with her life. She had not yet made peace with her life!

There was a part of her deep down that knew she needed to find her way to me. She said herself, "I didn't have a name, I didn't have a face, but I was looking for soul signature." She was looking for an energetic pattern and that's where Aria came in.

We do not create our tapestry by ourselves. We are not the only ones holding the yarn. Other people come along with different color yarns. They weave it into our warp and it becomes a beautiful image or a black shadow. With each new yarn of bright beautiful color, or the dark shadow in that warp, in that color of yarn; those are points to continue with the dark or to move forward with joy and flow with the color. It is up to us to keep choosing people and opportunities that enhance our tapestry, that make us stronger and better, and that make us more of our soul essence.

Aria brought her beautiful colored yarn to Grace and together they brought their threads to my tapestry. Together we created a beautiful picture.

Aria helped facilitate Grace and me finding each other. I was wholly in the flow of allowing that to unfold in the most natural way. I knew in the depth of my belly the pregnant cat that had been hit by a car was my cat and I would adopt one of her kittens. I didn't need a photograph; I needed nothing. But I did need her to have her kittens so that I could bring Grace and Aria home.

I remember the first time I saw Grace. She was a shy cat. I have no idea why, but I was a little afraid of her.

There was awe of her beauty, or maybe I just understood at some level the trauma she had gone through. I have always been cognizant of other people's pain and not wanting to intrude, but wanting to be respectful. I have always given people their space. So perhaps there was a little bit of that with the cat.

I do remember seeking her out downstairs at the foster home. She went downstairs when all the kittens went upstairs! She was all by herself and I can still feel the heaviness. Sorrow? Pain? Injury? Exhaustion? She was more hurt than we ever grasped. I dared to touch her; I dared to pet her. I thanked her for having the kittens and bringing one to me. I thanked her for bringing me Aria.

Your Turn

The weft and the warp.

Weft = the threads that weave our life experiences into our tapestry.

Warp = Divine perfection.

I invite you to take a look at your life up to this moment. Who has brought you colorful threads? Who has brought you dark colored threads?

Look at the pictures in your mind's eye. Choose one to focus on. Who brought you the colorful threads? Feel that person or group of people. Feel the experience. What does it feel like? Do you find yourself smiling? Do you feel your eyes light up? Who is around you? Do you feel joy and gratitude? Maybe you feel endless happiness?

Write about it. Take note of anything that comes up. As you write you may find you remember more details.

I challenge you to write that person a note of gratitude. Send it in the mail. Make it tangible.

You may find yourself being drawn to the dark colored threads. Perhaps it's a painful memory or experience that needs healing. Looking at the pain full on will help with the healing. What does it feel like? What happened that hurt? Move past the need to place blame. Seek grace. Seek forgiveness.

Write about it. Emote about it. Seek help if it's required. Seek healing. Make healing your priority over the need to be right.

Write With Grace

Journey
To
Grace

CHAPTER X

Fear

Take the hand of grace knowing you are
always supported.

Several months after Grace's death, she shared some poignant thoughts. She said to me, "My last few days with you were for me to work through pain and fear, my fear of death and dying, my fear of leaving the body and the physical world. I needed to work through my fears of being alive.

Bodies have never been a good thing for me. They give out. I tend to pick bodies that give out. In all lifetimes. In all time and all space. I have not chosen a body that worked for me long term because I was too afraid to live in them.

We all have these fears. There's not a single being that doesn't fear being where they are. We are all connected on that level. It's not a humanity thing, it's about being "Beings." Insects have fear about being in an insect body and bears have fear about being in a bear body. Humans have these same fears. We need to give everybody grace for being in their bodies."

Armadillos roll up in a ball and become still when they are threatened, this is their survival mechanism. It serves them well. But we are more evolved and we have the option of choice. The choice I would like for *you* to make is to drop into your heart and choose an action from a place of love and grace. Then trust that the decision you made from love and grace will take you down the path to another decision, which will take you down another path, which will take you, etc., etc. This is what my Grace referred to as taking one paw step forward.

When we are afraid, the natural inclination or body response is to stop, hold your breath, and tense up. When you do these things you are not moving forward, you are literally stopped. Your muscles are contracted and you are not moving. By holding your breath, your lungs are no longer moving. You are immobile. The natural tendency to stop and be still does not support us. When fear grips your heart it is important to take a breath, stop, take another breath, focus on your heart center, and make a decision from that place.

Fear takes us to our head, to that part of the brain that has been conditioned for century upon century. It takes us to our natural flight or fight response. To have success and forward movement we must start from a state of calm.

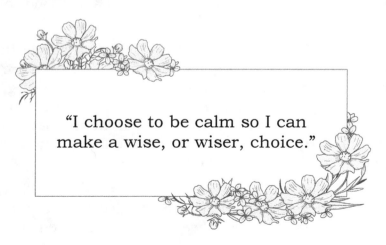

"I choose to be calm so I can
make a wise, or wiser, choice."

The wiser choice, the choice filled with grace as we are facing our fears full on, takes that weft and changes the color from the black which could be a dark shadow and into a colorful thread. It now has the potential to create something of beauty. I am not discounting the darkness that we all experience, but by choosing to face the fear from a place of grace, by choosing to face fear and take a positive action, we are changing the color of that thread. We are changing it to a beautiful color to create another amazing image in the tapestry of our life.

When we come to the end of our life and we are given a glimpse, an overall picture, of what the tapestry we have woven looks like, then we review our choices. When doing so we then see a plethora of beautiful scenes that are colorful, share our story, and show us we have lived a beautiful, colorful, rich, love and grace filled life. The beautiful vignettes that we weave into our personal tapestry are something to behold. They reveal where we have made the intentional choice to drop to the heart and by dropping to the heart we chose love, hope, and grace.

We aren't always faced with things that are scary. Sometimes it is an unexpected fear. Something grips our heart and we don't know how to mobilize. We don't know how to move forward because we are catatonic due to what has happened. Perhaps you have received unexpected news, been given a pink slip at work, or the man you have been married to for thirty-five years decides he's done and wants out. When life is threatened, being afraid is a very natural response.

When decisions were being made on Grace's behalf, she trusted that each step, each choice would get her further along to eventually meet me.

So what is your goal? Where do you want to end up?

Every choice takes you further away or closer to that goal. When fear grips your heart and comes crashing down on you at once, give yourself permission to sit with those fears. Don't do it very long, but be with the emotions. Limit your time and don't give in to depression or defeat. Don't allow fear to win, but at the same time honor the feelings of fear. Feel the fears and move through them. Take the hand of grace knowing you are always supported, always provided for, and always have a guiding force beside you. It's your job to call upon grace.

How else can you move through fear with grace? With trust. Trust in something bigger than yourself. Trust will give you hope. When you have hope you can move forward. On the other side is what you need. Love. Freedom. A breath of fresh air.

Grace said it best when she said, "I was afraid. I couldn't trust anyone. I was afraid! And I had kittens! Now what? I had to move through my fear of not trusting so we could be taken care of and find the love we really needed."

Your Turn

How do we choose grace? How do we actually *choose* grace? Here are some ideas:

- Close your eyes and take a deep breath. Allow your shoulders to relax.

- Soften your eyelids. Take another deep breath and exhale loudly.

Is that enough? Is that enough for you to calm down and drop into your heart? Maybe you need more than that. Here are some other options:

- Add movement.

- Add a qigong or heart opening exercise from yoga.

- Dance.

- Take a walk, go running.

- Add aromatherapy to anything that has been already suggested.

Many times when we fear something we live our life from a place of complacency. We allow life to happen *to* us. We forget we have choices. We always have choices. We make many choices throughout our day, in fact every moment we are making choices. Here are some examples of choices we make every day:

- What side of the bed do I get up on?

- What do I want to wear today?

- Do I want to bathe in the morning or before bed? Or both?

- What's for breakfast?

- Do I take that phone call?

- Do I have to go to work today?

- If you live with others (husband/wife, partners, roommates), who does what?

Try this —

Young Living Essential Oils has a blend called *Harmony*. Put a drop in your hand, rub your hands together, create a scent tent, and take a deep breath. This not only engages the breath, it involves the brain. The physiology of your brain will be impacted. Or do the same thing with the *Harmony Essential Oil Blend* and your heart opening movement exercise.

Write With Grace

CHAPTER XI

Letting Go

*Continuous acts of surrender are necessary
to live a life of grace.*

In the distance I noticed a huge oak leaf
as it fell with such grace and floated down
into the creek. The oak leaves were huge,
larger than an adult hand. This particular
leaf was cupped, resembling a boat. It
was as if I tuned in to the essence of the
leaf, or the tree, as it released the leaf.
It felt like a suspension of time and the
only thing that mattered was to witness
its descent. It felt sacred, pure, utterly
peaceful, and full of grace.

W hat if struggle was intended to be like the fall of this oak leaf? What if letting go looked and felt sacred? Pure? Peaceful? Indeed, what if it felt like grace? What if letting go was just that — letting go of something, someone, an idea, and a way of being— and not meant to be hard and emotionally painful?

Those moments when we are so present and in the "now," that is where time stands still and wisdom lives. This is when we can easily make decisions without agonizing thought because we decide from a place of power and strength.

"Now" is where possibilities live. "Now" is where fear is not.

How can change and struggle be handled with grace? See it for what it is. Resistance changes nothing for the positive; it makes the issue worse. Face it full on. Look at it and really get to know what the situation is asking of you. You may find your thoughts were completely unwarranted and erroneous. Simply realizing this can diffuse it and change suddenly becomes easy. Honor the struggle for the gift it has for you.

Recall the pearl. The coveted pearl is the result of an irritant within the oyster's shell. The struggle inside that oyster created a lovely pearl that many women desire. What if the tension within your struggle resulted in something of equal beauty?

When you both honor and fully see the struggle, you can then use it as a leverage point. Leverage yourself out of the undesirable situation on purpose. Intend to go forward and step into a newer version of you.

Grace reminds me often, "We are always in a state of letting go!"

What happens when you are holding a helium balloon? The balloon in its natural state, struggles on the end of the string. It does that because it knows it is made for other things. It is not meant to be tethered to a string. It knows it is made for something bigger, something faster, and something that can go higher. It knows. It struggles on the end of the string because it has a life and a mind of its own. It wants to be soaring in the sky. It wants upward movement, but we hold on to the string keeping it close to the ground.

What happens when we let go of the string? Either on purpose to watch it soar or because something distracts us and we accidentally let go. We watch it soar. Children will "ooh and ah" in wonder. We on the other hand, fret. How did that happen? How does it soar like that?

Our natural tendency is to want to soar. Our soul knows we are made for greatness, our soul knows we *are* greatness. Own your greatness. Own it.

Why do we struggle so? Why do we struggle within ourselves? Why do we shy away from shining brightly?

Letting go is sometimes conscious, intentional, or on purpose. Sometimes it is necessary because holding on is more painful than letting go. Sometimes it just happens as we toss up our hands and forget we really wanted to hold on.

A Lesson on Letting Go from Grace

Grace has helped me learn to let go. She often reminds me that, "At any given moment, at any given fork in the road, letting go is required. You cannot move down any new path without letting go of where you are. Movement of any sort requires letting go. Letting go of the current now to get to the new now. Truth be told, we are always in a state of letting go. Sometimes letting go feels bigger for some things than others. And indeed it is.

For you, dear Mama, to write with me and hear these truths, you have to let go of other options and other activities in which to engage. You choose to flow and commune with me and to allow the wisdom to come forth in this manner. You are letting go not only of other activities, but how you communicate and how it should look.

You allow the flow."

When we look at fear head on, we let go by allowing something different and new to appear. We have an idea of our intended direction — that's important — and our openness allows the weaving of our colorful tapestry.

Continuous acts of surrender are necessary to live a life of grace.

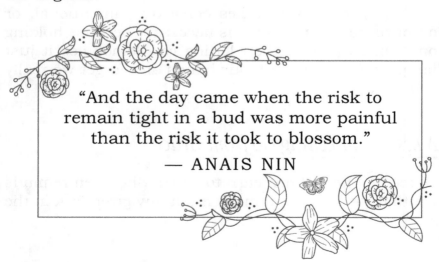

"And the day came when the risk to remain tight in a bud was more painful than the risk it took to blossom."

— ANAIS NIN

Your Turn

What are you in the midst of letting go of? Are you changing jobs or career? Are you letting go of a relationship that has gone stale and no longer makes your heart sing? Is it a habit that every time you do it, you find yourself wondering why?

Identify one thing that you could actively look at and practice the art of letting go. Envision it as a jacket. Practice taking it off and experience how it feels.

🐾 What would it feel like to get up in the morning and go to a job that truly satisfies you?

🐾 Or visualize yourself giving notice at work and visualize the new opportunity you are stepping into.

🐾 What if you let go of that person you consider a friend, yet deep down you know it's not true.

🐾 What if next time you find yourself doing your habit that leaves you wondering why, you stop — right in the midst of it — and experience life in that moment. Take off the jacket and feel the nakedness. Breathe.

At this point you have two choices: drop the jacket onto the floor and walk away or put it on and promise yourself you'll try again tomorrow.

Write With Grace

CHAPTER XII

Other's Grace

*Grace from others helps us take those
paw steps forward.*

Letting go is not a solitary event. It is a tiny, bit by bit, process. Grace calls it a paw step forward. Any forward movement is a success. Grace from others helps us take those paw steps forward.

An hour before Grace was to be put down, a dear friend came over to say goodbye. He is very uncomfortable with death and I could feel his discomfort.

He observed me interacting with Grace, giving her love, strokes, and speaking to her soothingly. He had a look of pain on his face. His look suggested immobilization which we talked about earlier. He didn't know what to do; he didn't know what was expected of him. Not that I had expectations of him, but what does one do in a situation like this? Death is something we don't learn about. We are typically shielded from it.

The look on my friend's face melted me. The look shone light on the strength I had had for many days in order to be there for my Grace. I fell into his outstretched arms and sobbed out the pain of her illness, the pain of knowing that in a matter of minutes I would be leaving with her and coming home without her. All of that exited my body with my tears.

There was a moment in that sobbing embrace where I felt a shift within me. I felt a moment of time suspended. It was Peace. It was Calm. It was Now. I knew it was a magical moment of healing, for my friend is a powerful healer.

There are two types of grace from others: the grace we offer someone who is struggling through a difficult life episode and the grace that supports and guides through a decision or transition.

My friend visiting to say goodbye is grace through a difficult life episode. The holding space, being there, not needing to use words; these are hugely important during hard times in life.

Aria, her grace in my life, in her mother's life, was to ensure that I follow the signs from one choice to the next that led me to the shelter where Grace was. Think of the orchestration of that!

That is grace. That is manifestation. That is God.

When you think about how this soul which is Aria not yet incarnate selected not only her mama that would give her birth, but she selected her human mama that would take them both in, that is divinity.

I took them in as my kitties. I never had my radar out for two cats, but when the idea presented itself there was never a hesitation. It was a knowing, a feeling. That was Aria. That was her grace for both Grace and me, as well as for herself. And it was also my allowing. And into my tapestry was woven something more beautiful because of it.

I've given you two stories. One where the grace of others helps someone through pain and the other a story of manifestation that is for the highest good of everyone involved. The grace of others.

Who are you here to serve? What are you wanting or being asked to manifest?

Acting upon that is grace. Had my friend not come to say goodbye to Grace, had he not done his magical healing that he doesn't even know he did, would I have survived? Absolutely. Would things have been different? Of course. Would I have processed the death of this beautiful feline that has taught me so much? Yes. But the gift that he gave me is rather profound. He showed up. And his thread was woven into my tapestry.

Your Turn

Recall a time where you allowed someone to bring grace to your own struggle. Remember how that individual served you. How well did you receive it? Did you acknowledge how valued this grace was? If appropriate, give that person a call or send a note of gratitude.

Is there a time when you brought grace to someone else's life? Was it a friend? Or perhaps someone who was in danger and you stepped up and into their life?

Feel the feelings. Remember the memories. Reflect.

Write With Grace

CHAPTER XIII

Trust

*It was only at the very end of my life that
I could see the Truth.*

Grace struggled with trust. She once said to me,
"Trust — that was the biggest lesson of my time on
earth. Trust. If I had not trusted and moved through
my fear and dealt with my own traumas, fear, and pain
— trust me that was not easy — I would not be writing
this book. I would not have a voice. I would have stayed
quiet. I lived in the lie of my story. I believed my first
humans. I believed them! If I had you in the beginning,
Mama, my life and my belief would have been so very
different. I know this.

When I was with you I couldn't fully understand and
appreciate all you did for Aria and me. You cared. You
care! You love so deeply! Mama! You really did love me,
and I couldn't let myself love you back, not until I was
scared and weak and dying. Then my love for you burst
out of my breaking body. Boom! It started gradually and
then burst open. And I would have moments of abject
affection for you.

I make you cry again. Love hurts. Maybe that's why
I walled off my heart. Love and trust hurt. By the time
you found me I had nothing left. All my energy went into

healing my own trauma and I had kittens to nurture too. I was so broken. Then they gave me three more kittens to love and care for.

Only when you brought me home did I have time for healing me — myself. But I was scared, confused, and distrusting. I didn't feel safe. And I know I misbehaved and that certainly didn't help."

Grace eventually came to a point where she was so sick that she relinquished control and allowed me to love and serve her. She allowed me to give her grace.

To allow someone to serve you, to love you, and to help you is a precursor to trust. Or is it?!

Trust, allowing, and vulnerability, they are a loop. What comes first? What's needed first? They all need to happen in order for anything to happen at all. There really isn't an order. Linear, sequential thinking is of this world and we are discussing higher level thinking and ways of being.

The Crescendo

In music, as when an orchestra plays, there's a build up to the crescendo. The strings, the percussion, and the wind instruments all come together in a loud climax.

That's what this trust/allowing/vulnerability discussion is, for they cannot be separated. Each plays their individual role with individual notes and they are all required and necessary to create the desired outcome. They are all required for the whole experience to come together and be meaningful. You need the love. You need the allowance. You need the care and understanding.

You need the vulnerability. You need the gratitude and service. You need the receiving. All are crucial for the crescendo of healing.

In order for someone to trust, they have to allow something. To allow something, they need to be vulnerable. That is what Grace struggled with the most, the vulnerability. She could not trust or relinquish control until she was so broken there was nothing else she could do.

When you allow someone to help you, you are being vulnerable and you are trusting in that person's sincerity, ability, help, understanding, love, and caring.

As her body deteriorated, Grace was finally at a point where she felt safer in my arms than on the bed or the floor.

As I think back to when I took her to the hospital, not having any idea how sick she was, I think of the long wait. The Virgo in me assumed I would be going to the hospital, dropping her off, and they would do whatever they were going to do and it would all be fine. It never occurred to me that I would be waiting for two hours. I now look back and I see the healing that took place in those two hours. In that waiting room, for the most part, she allowed me to hold her. She didn't struggle. She allowed. She acquiesced. She acquiesced into my arms. I could feel her surrender into my body.

I am so grateful that she trusted me to be there for her. That's all I could do at the end. All I could do was *be* there for her. I barely slept. I was there for her. I could not rest. I was invested in her well-being. I just needed to be there. To just be. I wanted her near me. I didn't want her to suffer alone.

"It was only at the very end of my life, Mama, that I could see the truth," Grace said.

Near the end of her life Grace had the vantage point of the Tapestry. She was removed from her third dimensional existence and she saw how I brought bright, beautiful colors to her tapestry. She saw that she was loved and that the colors of love are beautiful colors.

Grace told me she gained new beliefs during this time. She explained by saying, "I misjudged you. I was so cloudy with hurt and pain that I didn't trust. I didn't know how to trust. I didn't believe I could trust again. I didn't know what trust was.

Near the end, though, you showed me what trust was. You demonstrated and showed it all to me. I knew you were there for me. I knew you could be trusted to have my best and highest interest in mind and at heart. You could easily have put me down that snowy Sunday afternoon. Easily. Why didn't you?? Because you loved me. Was that the beginning of my trusting you? Yes. Yes! You chose the hard choice for you.

I am so glad you had Kristen talk for me. I like Kristen.

Yes, you made the choice that was best for me and hardest for you. I love you for that.

Thank you.

Thank you.

Thank you."

Your Turn

In what area of your life are you being asked to trust? Examine your emotions; how does your body feel? Are you resistant? Is it easy? Challenging? What's holding you back? Can you identify where you are in the symphony of trust/allowing/vulnerability? Who can you go to that will support you?

Write With Grace

CHAPTER XIV

Grounded in Grace

Being grounded in grace is giving up control and not having to always know what's next. 🐻

Be grounded in grace. Embodying your divinity is grounded grace. Grace feels like an exhale. It feels soft and gentle. It feels nurturing, like a caress. Yet grace isn't always pinks and purples and sunshine. Sometimes grace requires grit, but you can be grounded no matter what type of grace you're experiencing. It requires learning how to listen to your body and how to be in awareness.

Grounded grace comes with complete alignment from the top of your head to the bottom of your feet. To be grounded is to be in your body, to be able to feel your body, to sense the energy flowing throughout your body, and to feel the expansion. That is full and complete alignment. Many people have difficulty with this. They feel from the head up, on a good day maybe from the neck up. They have no idea what's going on around them and no connection to what their body is doing. They are not grounded in any manner and they are certainly not grounded in grace.

How do you achieve complete alignment within your body? What does it take to be grounded in grace?

Awareness. Body awareness is what is required. When you are body aware, you are in your body. You can then bring your attention to the base of your spine or even lower. Then take the energy all the way down your legs and push it into the earth. If you have a hard time staying in your body, take it even further. Take the energy that you've pushed into the earth and take it down even more and tether it around a crystal in the earth's core.

Being gracefully grounded is:

Stillness

Assuredness

Confidence

Knowingness

Calmness

It is a knowingness of who you are, what you stand for, and where you're going next without the need to control the outcome.

Grace says, "Being grounded in grace is giving up control and not having to always know what's next. I didn't know. I didn't know if I was going to survive and pull through or if I would let go. I went back and forth. That is part of the grace. Not knowing is part of grace."

What are your values? It's important to know your personal values, your reason for being here, and who you are to serve, be it professionally, volunteering, or serving your family. Knowing with confidence is being grounded in grace. This isn't a difficult concept or thing to achieve. It's a simple concept. It is easy to achieve. It's about being authentic to who you are and not allowing anyone to tell you otherwise.

Learn to listen to your own Inner Knowing. Learn to authenticate your own truths through your sense of knowing. When you are confident in who you are and what you stand for (and what makes you happy and makes your heart sing, what makes you feel small, versus what makes you feel expansive) you have a pretty good sense of being gracefully grounded.

"Being grounded in grace is to know you are human and nothing is solid. Humans don't give themselves grace to be who they are. It hurts no one to be you. Humans are the worst at being who they really are. They put on masks. They think they have to be someone they are not. They don't give themselves grace and grounding to be who they are. The world needs fewer masks and more people like you.

When you are grounded in grace, you deal with your stuff because you are no longer afraid of things that shaped you into who you are. You can face things with fearlessness and courage, which then helps you dig deeper into who you are," explained Grace.

Being grounded in grace holds a confident knowing Sacredness. It is bringing the energy of the Divine down through the body and into the earth. That is what my Grace did.

Your Turn

Reflect on confidence. Do you recognize the voice that is your Inner Knowing? Do you hear the whispers? Are you in loving relationship with your own personal truths? Have you taken time to write them down and/or sit with them so you can embody them, be an example of them?

Your truths become the filter through which you make decisions.

Write With Grace

CHAPTER XV

Authentic Grace

The end of the resistance brings grace.

Authentic grace is to be real, genuine, and true in and to you.

We have moved through fear and we have learned to let go and surrender. We have talked about the importance of the grace of others in our lives as well as the role trust plays as we find our way to a state of grace. We have also looked at the importance and attributes of grounded grace.

Now we are in a place where our authentic self can shine.

"Letting go of who we no longer are is pivotal on our journey to authentic grace," Grace says.

Life is a journey of uncovering our true self and the Divinity that we are.

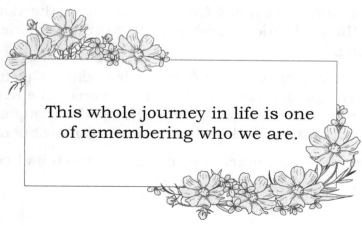

This whole journey in life is one of remembering who we are.

As we remember more who we are, we easily and naturally become our best and highest version of our authentic self. As we remember that we are a soul animating a body and begin to look at our life from the vantage point of the soul, everything changes. Our perspective changes on what holds true importance and what is trivial.

We are being asked to bring Sacredness into everyday life. This is Sacred Mundanity. While we are soul and sacred beings, we came to earth to experience what it is to be here.

Many of us go through a period of anger and anguish over choosing to come to earth and be human. It is a very real human struggle. When we are in that emotional state we really just want to go Home. It hurts so bad to be a person, to be a human, to have flesh and a beating heart and emotions. Until we *remember* we signed up for this! When we remember that we volunteered to come to this planet to learn a set of lessons and that we set it all up before we even found our body, it becomes easier. The end of the resistance brings grace.

For my Grace, that meant her death. When she was given the time and the space to go within and do her soul's work, she went through the fears, the doubts, and the realizations and she let go. She experienced healing.

In her struggle during those last days she finally realized how good her life was in our home. She realized how much she wanted to stay. Her body, though, was giving out little by little as her organs began to shut down.

I gave Grace, grace. I loved her as much as I could.

I offered her grace as I gave her space. I made the hard choice as I loved her and thanked her for letting me be her mama. And I told her it was okay to go.

I prayed for her to go without me having to take her to the clinic. But she held on. The little thing, she was determined that we both have that lesson and experience that was so holy and so sacred. The last gift she gave me was to die on my lap, a place she previously couldn't bring herself to trust to be. By the end she had gone through the fear, the letting go, the allowing of my grace in her process, and she was grounded in grace. She had embodied her name. She had become authentically, "Grace." And when she crossed over she became whole.

That is a love story. It's a story of love, allowance, grace, and commitment.

Grace sums this up well with her statement, "It's a story of Love. It's a path of Love. It's every step we take, every paw step forward, every choice, decision, and thought. Everything is part of that journey of living in Grace. It doesn't stop when you die. It never stops. Ever. It's a continuum. Whether it's a paw step forward (paw steps are smaller than human steps), whether it's a little step or a big step, it doesn't matter. It's part of the Journey."

Your Turn

Have you loved someone or something so much it hurt? Have you had to make the difficult choice to let someone or something go, whether by death or just parting ways? Have you felt anger and struggled with your choice to become human and come to earth?

Do you struggle to find authentic grace and be who you were meant to be? Sit quietly and think about who you are and who you are meant to be. Let go of the resistance and allow grace, sacred grace, in.

Write With Grace

CHAPTER XVI

A Veterinarian's Perspective on Grace

Sometimes a cat lives half of a life in nine lives, and sometimes they live nine lives in half of a life. ~ Angela

To have described them as specially matched, a unique pair, or bonded wouldn't have done them justice. When she was young and the babe was just a sparkle in her eye, she was hit by a truck.

People

Rescue

Surgery

One eye lost

But the baby was still there

After some time a soulful woman came along, welcoming her into a cozy warm home. The woman named her Grace and the veterinarian thought it was fitting, for grace had been bestowed on her in so many ways. She was trying to move gracefully through life even though it was a challenge.

Some say they are just cats.

The day the veterinarian saw the mother and baby reunited with visible knowing and affection, she knew they were not just cats. The sparkle of spirit was there. To have described them as specially matched, a unique pair, or bonded wouldn't have done them justice.

Aria came into life like a whirlwind. Her name was given to her by another in a dream, and some say that her name means "lion of God." Her sweetness was matched only by her mother's and for that she was grateful.

Then one day Grace became ill. The emotion, the trauma, and the worry all built up in her body. From the veterinarian's perspective these things combined dealt a fatal blow by accumulating and developing into a tumor. She tried to tell the soulful woman that she didn't feel well.

Pee on the rug

Tell the lady that talks to animals

Pee on the rug again

Tell someone, anyone

And they tried to help her. The people tried to rescue her again, but this time her body had enough and she made her peace in her heart. Sometimes a cat lives half of a life in nine lives and sometimes they live nine lives in half of a life. She was young, but when she succumbed she had lived a full nine lives filled with adventure, trauma, solitude, motherhood, family, and ultimately love. A mandala of love was all she could share with the soulful woman to express what she meant to her. And she knew that Aria was in good hands.

And they say they are just cats.

Signed,

Angela

CHAPTER XVII

Aria's Grief and Healing

If there's anything I've learned from being a mama to kitties, it's that they are "human."

One month since Grace's death, Aria still mourns her mama. Since Grace's acute and unexpected illness and death, Aria has had many different cries. Cries I never heard before; cries I have not heard since. Yet it is obvious she mourns because she wanders the house singing a mournful song, usually in the morning.

After Grace died I bought special toys for Aria to give her play time and attention. She did the most curious thing each evening; she brought her favorite toy to me at 8:00 PM, pretty much without fail. When the time changed, she was an hour off! I mentioned this to a friend and she said Aria didn't get the memo about the time change.

If there's anything I've learned from being a mama to kitties, it's that they are "human." They have distinct personalities, ways of being, and preferences. Just like we do.

As I share Aria's grief with others, many suggest I simply get another cat to keep her company. I know they mean well, but if a human in my family died would the suggestion be, "Get yourself another husband, mother, or cousin?" Probably not.

Ultimately everything is about consciousness. All life revolves around a whole perspective on life. What did the cat come here to do? What was she to learn? Teach? Contribute? Each soul (human, feline, or otherwise) has a role to fulfill and lessons to impart and learn.

Maybe Aria's voice is to help people value the importance of feeling and living through the grief? I've been teaching for years and have found the only way out is to go through. There is no more going around or skirting the issue. Try it. Eventually it comes back and it's bigger, demanding attention. *Through* is the only way, even though it may be uncomfortable.

Thirteen months after Grace left us my daughter had a destination beach wedding which led to me being away from Aria for nine days. I made sure she had the care needed while I was gone, as this was the longest I had ever been away from her.

Aria really missed me and could not have enough snuggles upon my return. I made sure to give her the attention and reassurance she needed. Since that time, her playfulness and eagerness to play and engage has increased.

Since my nine day absence for my daughter's wedding, Aria brings toys to me not necessarily on any schedule, but out of a sheer desire to play. I come home to find she has brought toys up to the living room or I find toys on the stairs that didn't quite make it all the way. She communicates via chortling and she engages me by looking deep in my eyes and nodding her head.

She chortles and I wonder what she's trying to say — really — but I know she wants attention and to be seen.

Aria regularly lays on her back asking for her belly to be rubbed, very similar to what a dog might do. Sometimes I think she's more human than feline and sometimes I think she's got a little canine in her!

Her grief over her mama's death took months for her to process in her own way. I noticed after about six months she seemed to settle into the new way of living alone with her human (me!). Kristen tells me Grace sometimes pops in to visit. Indeed, I have seen her shadow. When people came to visit after about that six to seven month mark, they noticed Aria was different. Her energy wasn't as heavy and she seemed brighter. She is now more at ease and peace.

"We have all but muffled and muted them (the sounds of grief) in humanity - sometimes it is the animals that teach us how to be human again - through the bridge of pain ... and it is hard."

— ROXANNA L. RUTTER

My Aria is extremely bright. There are some days where she chooses to be very solitary and curl up on the massage table which always has the BioMat turned on. It is downstairs so she is alone. Periodically I will go and make sure she's okay, give her a pet and a kiss, and go back to my day. Sometimes I say, "I really like it when you are near me." Interestingly, she will then come upstairs and curl up under a chair near me.

She loves to snuggle on my shoulder. I adore and welcome this. We have settled into the newness of our relationship without Grace. Kristen tells me Aria doesn't understand all the fuss about the book Grace and I are writing. But I believe she loves the idea of her mama living on in a book that touches your heart; for she knows it will. She knows it has. Grace has left her paw prints upon your heart.

Your Turn

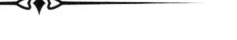

Have you lost a pet at a time when you had one or more pets still living? This is a unique situation that most have not experienced. It certainly is not openly discussed — for this grief is unfamiliar to many of us.

If you find yourself in this situation, honor the loss. It's different. It's different because "life goes on" and you still need to be there for the family animal(s) that remain.

Write about your experience. If you have a spouse, roommate, and/or children, they factor into the loss and experience of the death, too. Write. Allow yourself to grieve. Doing so begins true healing.

Write With Grace

EPILOGUE

How I Supported Grace During Her Illness

You may find yourself in a similar situation and wish to know more specifically how I supported Grace. If so, this section may be for you.

I relied on Kristen to be Grace's voice. My emotions and energy levels were such that I needed the truth and certainly did not need the pressure of guessing. I cannot stress enough the value an animal communicator brings to the relationship between the human and animal.

I listened to what Grace wanted. This was not about me or taking the easy way out. It was about her soul's journey.

Music soothed and supported my Grace.

The first time I played the piano after bringing Grace home I felt something on my shoulder. I turned around and Grace was reaching over from a nearby table tapping me on the shoulder. Needless to say, I was astonished at this behavior. Grace loved music. It spoke to her deeply and she responded so readily that I used soft, gentle music in the background to sooth her.

There were other times when I would play a newfound song and she would excitedly jump up and sit beside the music source. I believe the music was soothing to her brain as well as her spirit.

Lee Harris has a piece of music called, *Adventures in Sound,* a sound toning in three parts: *Expand, Integrate, and Open.* It is intended to do just that: to expand, integrate, and open the listener. One morning as I was playing it, Grace responded demonstratively toward it. I then played it more frequently to support her healing.

Aromatherapy

I diffused vetiver essential oil. Actually, both cats asked me to diffuse that specific oil! It has a deep, woodsy, grounding aroma and it helps with stress and emotional trauma. And yes, it is safe to use with cats (and other animals). I use only *Young Living Essential* oils, known for their purity and efficacy. *

What is a "scent tent?" After you have placed a few drops of essential oils into your hands, place them over your nose and breathe deeply. That's it! Easy!

Amethyst BioMat

I had the Amethyst BioMat on and loved seeing Grace lay on it, knowing it would help her condition and provide her comfort. This mat emanates a soft, gentle, nurturing heat that animals and humans love! *

* See Resources page for more information.

Crystals

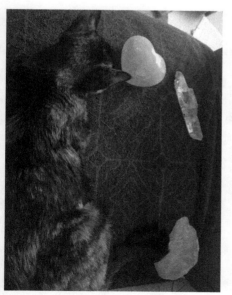

I placed crystals around her to support her energetically.

These photos were taken thirty-four hours before Grace's death and less than two minutes apart. Crystals are powerful healing medicine. Under the towel is the Amethyst BioMat.

Meet Kristen Scanlon

Kristen muses over her time spent talking with Grace:

"Whenever she talked,
I was in awe of her wisdom."

Kristen Scanlon is an exciting and creative woman who loves working with both humans and animals alike. She has been a professional animal communicator, BodyTalk practitioner, and Access Consciousness practitioner in the Twin Cities, Minnesota area since 2008.

Kristen learned she had the ability to hear and converse with animals through a journey of heartbreak, skepticism, and persistence while pushing past the conventional ideas of what it meant to heal after years of abuse and trauma. She was persistent in wanting to know how to lead with her heart. She learned through the love and grace she received from the animals around her. She breaks through the barriers between the animal and human world, creating a deeper awareness of what we all need in order to co-exist on the planet and *be* the connection. She brings forth the voice of the animals so you can hear their words of wisdom and guidance with not only their lives, but our own as well.

She has a knack for knowing what questions to ask to receive the important messages animals desire to share with you. The animals bring insight and clarity during the sessions she provides.

Learn more about Kristen at www.TalkPawsitive.com

Resources to Continue Your Journey to Grace

🐾 *Grace Notes with Grace:* A regular drop of "Inspirational Nourishment" in your inbox. Just what you need to align yourself with what is important to YOU. These messages will inspire confident action, nourish your soul, and create a sense of empowerment.

Get started today! www.GraceNotesWithGrace.com

🐾 *Lee Harris Music:* Grace responded to Lee's *Adventures in Sound: Integrate, Open, Expand* in quite a remarkable way. It's one of my personal favorites to listen to, especially after a major change/shift in my life.

Visit www.LeeHarrisEnergy.com—choose store, then music and you'll find it. You may also find other wonderful things that resonate with you!

🐾 *Ruff Start Rescue:* Ruff Start is the organization where Grace found herself. It is a "no kill" rescue based in Central Minnesota. They rescue stray, neglected, abandoned, and surrendered dogs and cats, as well as ferrets, guinea pigs, rabbits, and other animals.

www.RuffStartRescue.org

🐾 *Amethyst BioMat:* A high-quality infrared heating mat used throughout the world for healing and pain relief. Animals love the energy of this technology!

Visit www.MaritaInternational.com/resources for more information and to order yours! Grace sought out the BioMat.

🐾 *Young Living Essential Oils:* The world leader in aromatic, concentrated plant extracts carefully obtained through steam distillation, cold pressing, or resin tapping. Essential oils are used for supporting the body systems, emotional well-being and spiritual development.

Both Grace and Aria, via Kristen, specifically asked me to diffuse vetiver. Animals know what they need.

Visit www.MaritaInternational.com/resources to learn more and get started.

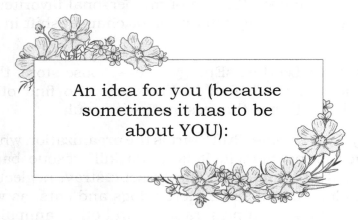

An idea for you (because sometimes it has to be about YOU):

Combining the BioMat with essential oils and intention has the potential to create profound experiences. Try it! (This is also another way of saying humans benefit from using the BioMat and essential oils!)

I am always innovating and creating new experiences. Look for what's new at: www.HarmonicWholeness.com

Acknowledgments and Gratitude

Three exceptional friends:

Beth, for believing there was a higher purpose for Grace coming into my home and for your encouragement to bring Kristen into the conversation. If you had not done so, Grace would have had a very different journey on this planet. And so would I.

Gabriel, for supporting me through the discomfort of death and bringing your healing magic to me when I most needed it. Thank you for honoring Grace in her last hour.

Nancy, for adoring and caring for Grace and her kitten, Aria. Your belief in our story is treasured.

Kristen Scanlon, animal extraordinaire. I could not have navigated the extreme emotions and uncertainty without you; thank you!

Angela, the best veterinarian ever. Thank you for your unique approach to veterinary medicine and bringing your loving honor for animals to Grace.

123

My Facebook audience who supported both Grace and me during Grace's illness and death, who encouraged and reached out to me, and who shared their stories of healing because I shared mine.

Jesse Krieger, my publisher, for seeing and believing in the value and gift of Grace's story and for helping the vision of this book come to life.

Kristin Davis, my editor, for making my words dance perfectly on the page.

It is quite humbling indeed, to realize what an impact this book is making — even before it is published. The support is overwhelming, much of it from unexpected pockets of people finding value in the voice I have given Grace. It is you — it is YOU — I am grateful for. It is the creative insight you have given me as to where and how I can take this work, this labor of love, into various places, and to various groups of people I would never have thought of. Truly, it takes a village. You are my village. You, dear reader, are part of my village and our story. Thank you.

About the Author

Marita Rahlenbeck has a natural way of bringing the Sacred into everyday life. She easily sees a higher vantage point, particularly during difficult life challenges. That's why, while it was not easy to watch her cat, Grace, suffer, she understood there was something much bigger at play. Empathy comes easy; as does nurturing.

Marita helps women feel fantastic in their own skin so they can shine from the inside out. While her holistic approach involves supporting the body as well as the spirit, she didn't always view life holistically. This happened over the course of three life changing events.

The first crack in Marita's protective armor started after a painful divorce and while transitioning from a stay at home mom to a single parent of a six-year-old. She first experienced "*White Angelica,*" an essential oil blend from Young Living Essential Oils, at an outdoor wedding. Shortly thereafter while at an expo, Marita visited a booth with many tiny bottles displayed. Of all the bottles to choose from, the woman speaking with Marita chose White Angelica for her to experience. Marita recognized this as a sign that these miracle drops were what she was looking for to provide emotional support during this difficult life experience. Marita's divorce was her catalyst into a spiritual opening and the oils helped with this, as White Angelica creates a gentle sense of wholeness within one's spiritual self.

The second crack was after a severe car accident resulting in a two and a half year journey to achieve a life without constant pain. Again, the essential oils provided physical and emotional support, as they are remarkably effective in supporting the physical body. After all, aromatherapy is the medicine of old.

The third crack resulted from a fall that left Marita with a traumatic brain injury. This injury left Marita broken and proved to be a life altering and very lonely recovery. After being told there was nothing they could do, Marita defied specialist's prognoses and determined to achieve optimal brain function without western medicine. Essential oils, among other modalities, again provided support on all levels — body, mind, and spirit.

The brain injury prevented Marita from being able to read, listen to music, enjoy educational material, and many other things, so she found new life in the kitchen. Her creativity was unleashed, and her love of food thrived.

Her recovery taught Marita how to "be." She learned the art of stillness and says, "I was on a journey to the mountaintop, without climbing the mountain."

These three life events bring what Marita stands for firmly into her life and make her sing. She found her "sacred" in aromatherapy, food, and the Sacred.

While it is true Marita is known for creating exceptional meals in her ridiculously tiny kitchen, she is also renowned for her aromatic and energetically welcoming home and group experiences. (She always smells amazing too!)

Just as an alchemist would, Marita easily and skillfully pulls three seemingly unrelated things together and fully brings them into her life *and* work. Marita is a master at not only meeting you where you are, but at asking penetrating questions to achieve results — results that YOU want.

Then there is Grace. Looking back, Marita sees the importance grace had in her life and how she values it. Her cat, Grace, solidified this within her core and life. Writing *Living with Grace* wasn't necessarily easy, but the acceptance and knowing it had to be written was. And in so doing, Grace was present and grace profoundly changed her. Listen and learn as Marita shares with you how to let grace, and Grace, into your life and soul. You may find your life profoundly changed, and even if not, you may just find grace. And we all need a little grace (Grace).

Marita and Aria live in Minneapolis, MN.

Every paw step forward is part of the journey of living with Grace.